CHAS. S. MILEY LEARNING RES. CENTER—LIBRARY
INDIAN RIVER COMMUNITY COLLEGE, FT. PIERCE, FL

Earth and Space

CONTENTS

The Big Black Beyond 2
Spaceship Earth 4
Cosmic Distances 6
The Sun and Stars 8
Stargazing 10
Our Solar System 12
Moon Watch 14
Flying Objects 16
An Astronomical Story............18
Blast-off! 20
Messages from Space 22
Space Hotels 24
Spacelabs 26
Future World 28
Cosmofacts 30

CREATIVE EDUCATION

The Big Black Beyond

The universe is absolutely everything that exists, including you. Mostly it consists of vast, empty, black space, which scientists call dark matter, but it also contains countless billions of stars, just some of which you see as twinkles in a night sky. How did it all begin? It is believed to have started with the biggest explosion ever, known as the Big Bang.

Is anybody here?

15 billion years ago the universe began

The Big Black Beyond

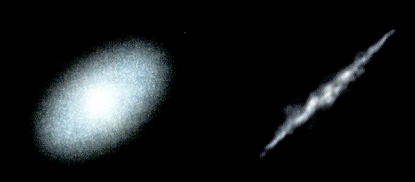

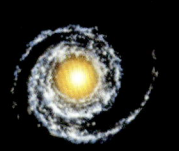

ELLIPTICAL GALAXY

BARRED SPIRAL GALAXY

IRREGULAR GALAXY

Stars and galaxies

Stars are big. Our Sun is a star, and it is bigger than a million Earths. Some stars are a thousand times bigger than the Sun. But all the stars together amount to only 0.00002 percent of the universe! Galaxies are huge clusters of about 100 billion stars, dust, and gas. They can be different shapes. Our Sun is in the Milky Way, which is a spiral galaxy.

The expanding universe

Since the Big Bang, about 15 billion years ago, the universe has been expanding and cooling. The galaxies are moving farther and farther apart. Maybe everything will continue to expand until the last star runs out of energy.

The big crunch?

One theory about the future of the universe is that it will stop expanding and collapse inward, so that all the galaxies crash into each other in a Big Crunch.

The dots on this balloon move apart as it is inflated. Galaxies do the same as the universe expands.

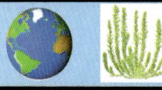

| 5 billion years ago Earth was formed. | 3.2 billion years ago first life forms appeared. | 230 million years ago the earliest dinosaurs lived. | 40,000 years ago human life began on Earth. |

Spaceship Earth

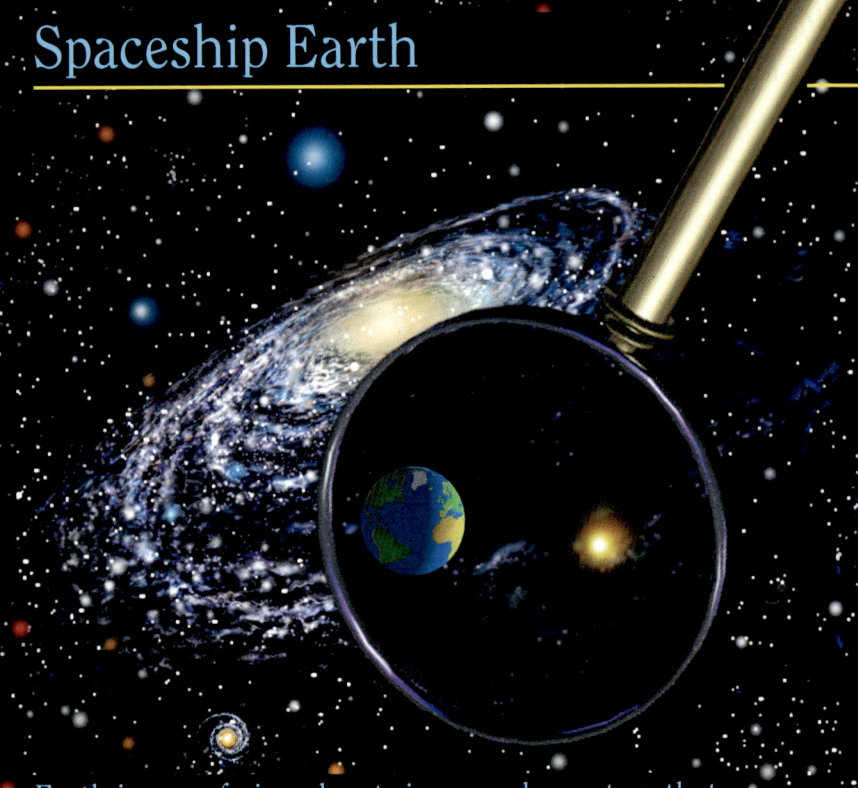

Earth is one of nine planets in our solar system that all move around a star called the Sun. There are about 500 billion other stars in our galaxy, which is known as the Milky Way. Scientists believe that there are a trillion galaxies in the universe. So if someone asks you for your full address, do you know what to say?

How the earth was formed

The Sun and the nine planets were made from a swirling cloud of dust and gas about 5 billion years ago.
The Sun formed in the middle of the dust cloud and cooler matter around it was shaped into planets. The leftovers became asteroids, meteors, and comets.

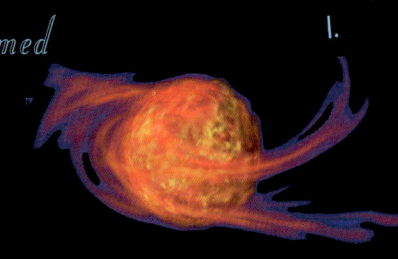

1.

1. 4.5 billion years ago, proto-Earth was a condensing cloud of gas. 2. 4 billion years ago the whole planet was very hot, molten rock

Spaceship Earth

Life on earth

Earth is unique in the solar system because it is the only planet able to support life. Other planets are wrapped in choking dust or poisonous gases and have no flowing water. Earth's atmosphere contains oxygen and two-thirds of the globe is covered by ocean.

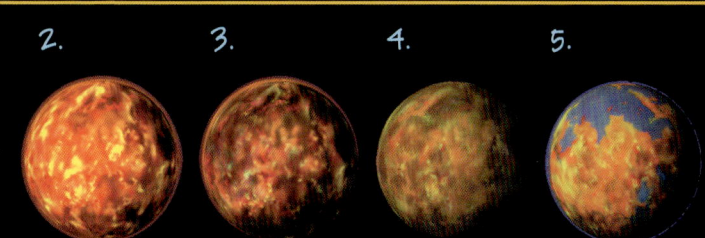

YOUR ADDRESS:
THE EARTH
THE SOLAR SYSTEM
MILKY WAY GALAXY
THE UNIVERSE

2. **3.** **4.** **5.**

with no atmosphere. **3.** Earth began to cool very slowly. **4.** Earth's crust cooled completely and an atmosphere began to surround the planet about 2 billion years ago. **5.** Oceans began to form on the planet about 1.5 billion years ago.

5

Cosmic Distances

When you switch on a light it shines instantly, or does it? In fact, it only seems instant because light travels so fast—about 186,000 miles (299,000km) a second! Distances in space are so enormous that we can't measure them in miles: We have to use light-years.

Traveling light
We don't know of anything that moves faster than light. In one year, light moves 5.8 trillion miles (9.4 trillion km). This is the unit of measurement we call one light-year.

Faraway stars
When you look at stars you are actually seeing the light that they give off. Light from the Sun takes 8 minutes to reach Earth. From our next nearest star, Proxima Centauri, light takes 4.2 years to reach us!

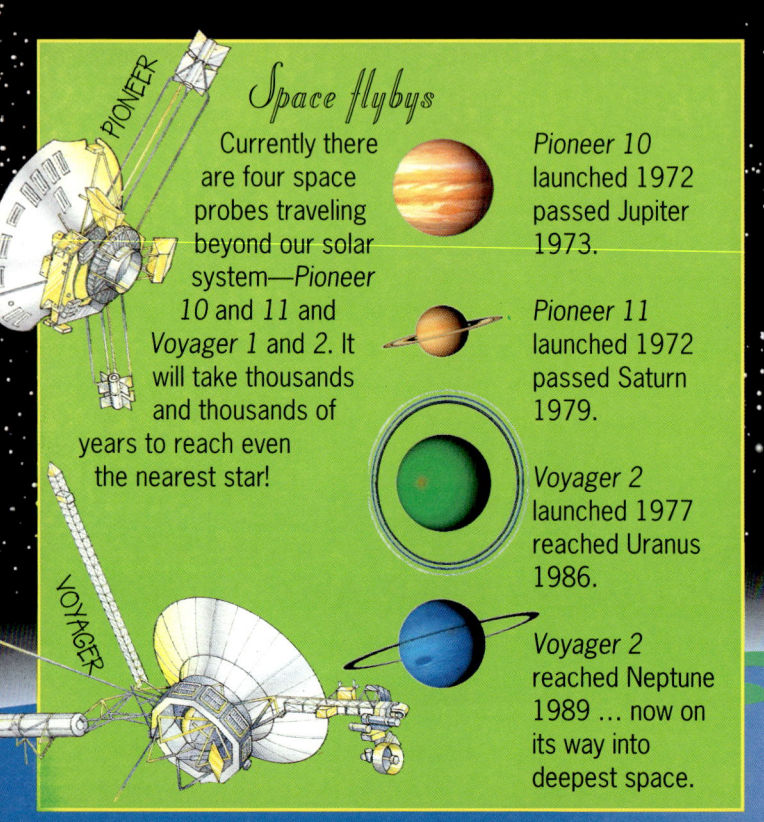

Space flybys

Currently there are four space probes traveling beyond our solar system—*Pioneer 10* and *11* and *Voyager 1* and *2*. It will take thousands and thousands of years to reach even the nearest star!

Pioneer 10 launched 1972 passed Jupiter 1973.

Pioneer 11 launched 1972 passed Saturn 1979.

Voyager 2 launched 1977 reached Uranus 1986.

Voyager 2 reached Neptune 1989 ... now on its way into deepest space.

Cosmic Distances

Light from the past

⭐ The Andromeda Galaxy is the most distant object visible in space with the naked eye. The light we can see left the stars in this galaxy 2.2 million years ago, when hairy mammoths roamed the earth.

⭐ Polaris, or the North Star, is 680 light-years away. So the light that we can see from this star left it during the Middle Ages, when knights wore armor.

⭐ Some of the stars in the constellation of Orion are 110 light-years away. The light from them left when the automobile was just invented.

The space shuttle travels fast. But it would take 219 days to reach the Sun.

The Sun and Stars

A star is a huge spinning ball of hot gas, usually a mixture of hydrogen and helium. The Sun is our nearest star and without its energy, there would be no life on Earth.

Solar flares shoot out from the Sun's surface and can reach far into space.

Heat energy can take 10 million years to reach the Sun's surface.

Heat energy is generated in the Sun's core where the temperature is 27 million°F (15 million°C).

Sunspots are dark patches of cooler gas. The smallest sunspots are 50 times the size of Africa.

The life and death of a star

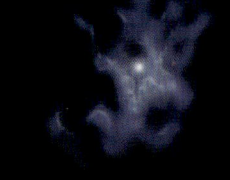

Stars are born inside clouds of dust and gas.

As the gases slowly compress a star forms.

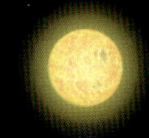

The star shines steadily for about 10 billion years.

The Sun and Stars

Solar energy
In sunny parts of the world, energy from the Sun's rays can be trapped by solar cells. It can then be used to create electricity for heating homes and powering equipment.

Black holes
Not all stars fade away quietly as white dwarfs. The biggest stars can explode in a blinding flash before collapsing into a core of matter that is so dense its gravity will suck in everything, even light. Scientists think there may be a giant black hole at the center of our galaxy.

WHITE DWARF

Then the star gradually cools and swells to become a red giant.

A red giant is very unstable, and as a result, gases from its surface blow off, leaving a small smoldering core of cinders. This core is known as a white dwarf and is all that remains of a star when it has completely run out of energy.

9

Stargazing

As our planet travels in its orbit through space, we can see stars in the night sky. The stars that we can see depend on the time of year and whether we are watching the sky from the northern hemisphere or the southern hemisphere. Can you see either of these constellations in your night sky?

THE BIG DIPPER
NORTHERN HEMISPHERE

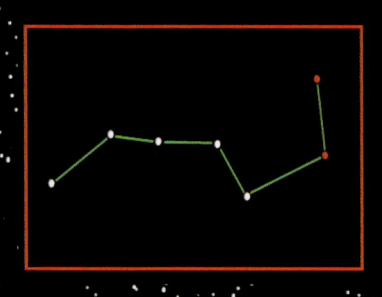

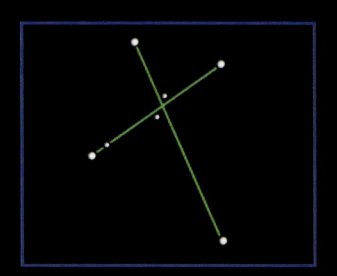

THE SOUTHERN CROSS
SOUTHERN HEMISPHERE

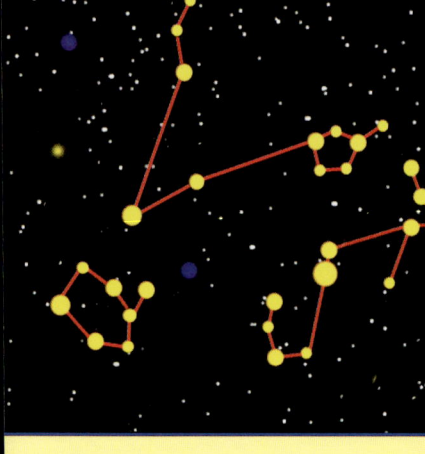

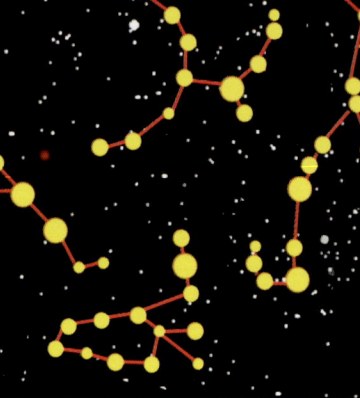

Pictures in the sky

People have always been fascinated by stars. The ancient Greeks and Romans grouped the stars of the northern hemisphere into patterns called constellations, which they named after people and animals from myths and legends. Twelve of these constellations are called the signs of the zodiac. Can you find them in this map of the night sky?

Aquarius

Pisces

Stargazing

Signpost stars

When the spaceship *Apollo 13* had technical problems on its lunar mission in 1970 and was forced to return to Earth, the crew had to fly the damaged spacecraft using the position of the stars as a guide.

Two stars in the Big Dipper constellation (marked in red on the opposite page) can be used to point to a very bright star called Polaris or the North Star. This star always lies directly above the North Pole and has been used by sailors for centuries to help them sail in the right direction.

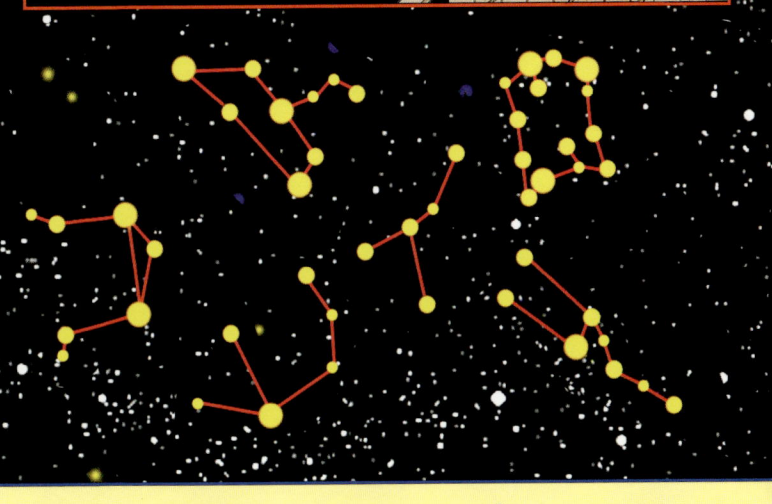

Aries	Gemini	Leo	Libra	Sagittarius
Taurus	Cancer	Virgo	Scorpio	Capricorn

Our Solar System

The Sun is at the center of our solar system. It is a massive ball of gases that produces as much energy as 100 billion hydrogen bombs exploding ... every second! The nine planets of the solar system and more than 60 moons orbit the Sun.

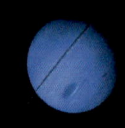

Our Solar System

The **Sun's** gravity keeps all the planets in the solar system in their orbits.

Mercury orbits the Sun in just 88 Earth days. It is a small, rocky planet scarred by craters and wrapped in a very thin atmosphere.

Venus is covered with swirling acid clouds, which trap the Sun's heat at furnacelike temperatures.

Earth looks like a bright blue-green jewel because it is covered by so much water.

Mars looks red because there is so much iron oxide dust in its atmosphere.

Jupiter, the largest of the planets, is a gigantic ball of gases with a family of 16 moons.

Saturn is a beautiful ringed planet. Despite having the chemicals to support life, it is icy cold.

Uranus is greenish in color because it spins in a fog of methane gas.

Neptune is a bluish globe of gas with hurricane-force winds and two narrow rings.

Pluto is the smallest, coldest, and farthest planet, and it has an irregular orbit.

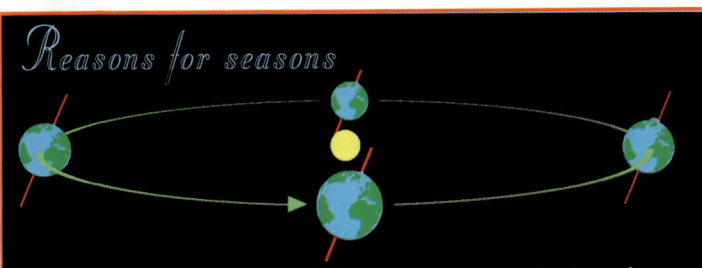

Reasons for seasons

As Earth spins on its axis and orbits the Sun, some parts of the globe are facing toward the Sun when others are facing away from it. When the northern hemisphere faces the Sun it has its summer. At the same time, the southern hemisphere faces away from the Sun and has its winter.

Moon Watch

The Moon is Earth's nearest neighbor and is only 238,857 miles (384,404km) away! It orbits our planet every 29.5 days. The Moon's surface is pitted by about 500,000 craters. These are scars left by massive meteorite storms that bombarded the Moon, which has no atmosphere to protect it, 3.8 billion years ago.

First man on the Moon
Neil Armstrong was the first human being to set foot on the Moon. He was one of three astronauts on the *Apollo 11* lunar mission. As he stepped onto the Moon's rocky surface on July 20, 1969, he said, "That's one small step for a man, one giant leap for mankind."

Moon motoring
Astronauts on the Apollo missions got around on the Moon using the lunar roving vehicle or "moon-buggy." It was battery powered and enabled crew members to travel farther afield to collect moon rock samples and take photographs.

Footprints forever
Footprints left by visiting lunar astronauts will not erode since there is no wind or water on the Moon. They should remain visible for at least 10 million years.

Moon Watch

New Moon    Crescent Moon    Half Moon    Full Moon

Phases of the Moon
Like all other planets, the Moon has no light of its own. It shines by reflecting light from the Sun. As the Moon orbits Earth, it seems to change shape because different parts of it are in shadow at different times. The diagrams above show the phases of the Moon in the first half of its orbit. These phases are reversed in the second half.

Tide times
The Moon's gravity has a pull on Earth that is shown by the daily patterns of the tides. Each time the ocean faces the Moon as Earth spins on its axis there is a high tide. When the ocean is not facing the Moon it will be at low tide.

Moon facts
The far side of the Moon, which never faces toward Earth, was photographed for the first time by the *Luna 3* spacecraft in 1959.

About 840 lb. (378kg) of moon rock has been brought back to Earth by the Apollo missions.

Temperatures on the Moon's equator can reach 240°F (116°C) at midday and then plummet to −260°F (−162°C) at night.

HIGH TIDE    LOW TIDE

Flying Objects

Planets, moons, and stars are not the only objects whizzing around in our galaxy. There are also countless lumps of rock mixed with ice, dust, and gas rushing through space: from tiny specks to huge boulders.

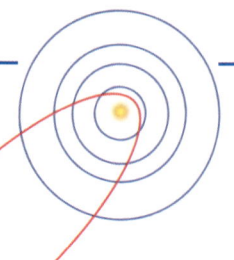

Some comets move around the Sun following a very elongated elliptical orbit.

What is a comet?
Comets are like enormous dirty snowballs. If they get too near the Sun its heat turns the ice to gas, which forms a bright tail stretching for millions of miles through space.

Meteoroids and meteors
Meteoroids are rocks that fall through space. If a meteoroid enters Earth's atmosphere, it starts to burn up and becomes a meteor or "shooting star." Sometimes they fall in meteor showers.

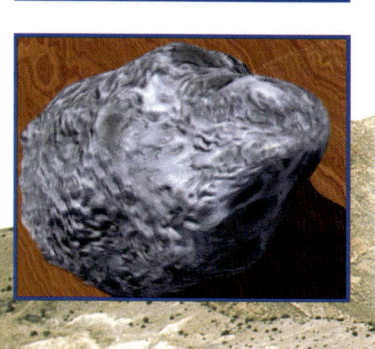

Meteorites
Meteors do not always burn up completely, and the rocks that survive their journey through Earth's atmosphere crash to the ground as meteorites.

Flying Objects

Come again comet

Halley's comet returns to Earth every 76 years. It was seen in 1066 and is pictured in the famous Bayeux Tapestry, which tells the story of the Battle of Hastings. The European unmanned spacecraft *Giotto* was launched in 1985 to make a close flyby of Halley's comet. In 1986, *Giotto* flew within 375 miles (600km) of the comet and took photographs of it.

Flying rocks

Asteroids are chunks of rock left over from when the solar system was formed. They circle the Sun with the planets. Most of them travel in a wide belt between the orbits of Mars and Jupiter.

Dino death

A massive asteroid fell to Earth in Mexico about 65 million years ago. The force of its impact caused tidal waves and earthquakes and sent huge clouds of dust into the atmosphere, disrupting the world's climates. Some scientists believe these dramatic changes may have caused dinosaurs to become extinct on our planet.

The Barringer Crater in Arizona is 570 ft. (171m) deep and 4,150 ft. (1,245m) across. It is the world's biggest meteorite crater.

An Astronomical Story

Long ago people thought that the earth was flat and that if you traveled far enough you would fall off the edge! They also believed that the Sun moved around Earth.

Galileo
Galileo invented the first astronomical telescope in 1609. It magnified objects only to 30 times their actual size, but Galileo used his telescope to prove Copernicus's theories.

Cosmic controversy
In the 1500s Copernicus theorized that the planets revolved around the Sun. His views went against the teachings of the church, so he did not publish his theory until he was lying on his deathbed in 1543.

Planet pioneers
Copernicus knew about the five planets nearest Earth. In 1781 William Herschel built a telescope in his garden and discovered Uranus, but Neptune wasn't identified until 1846 and Pluto not until 1930.

Lowell's martians?
In the 1880s Percival Lowell built a powerful refracting telescope in Arizona. He spent hours observing Mars and tried to prove that the channels he saw on the planet were in fact canals dug by intelligent beings.

KECK TELESCOPE

Today's telescopes
Galileo's telescope used lenses to collect light. Modern telescopes use mirrors. The Keck reflector in Hawaii uses 36 joined mirrors forming a surface 33 ft. (10m) across.

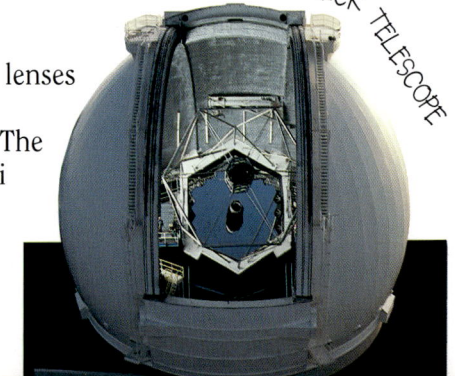

18

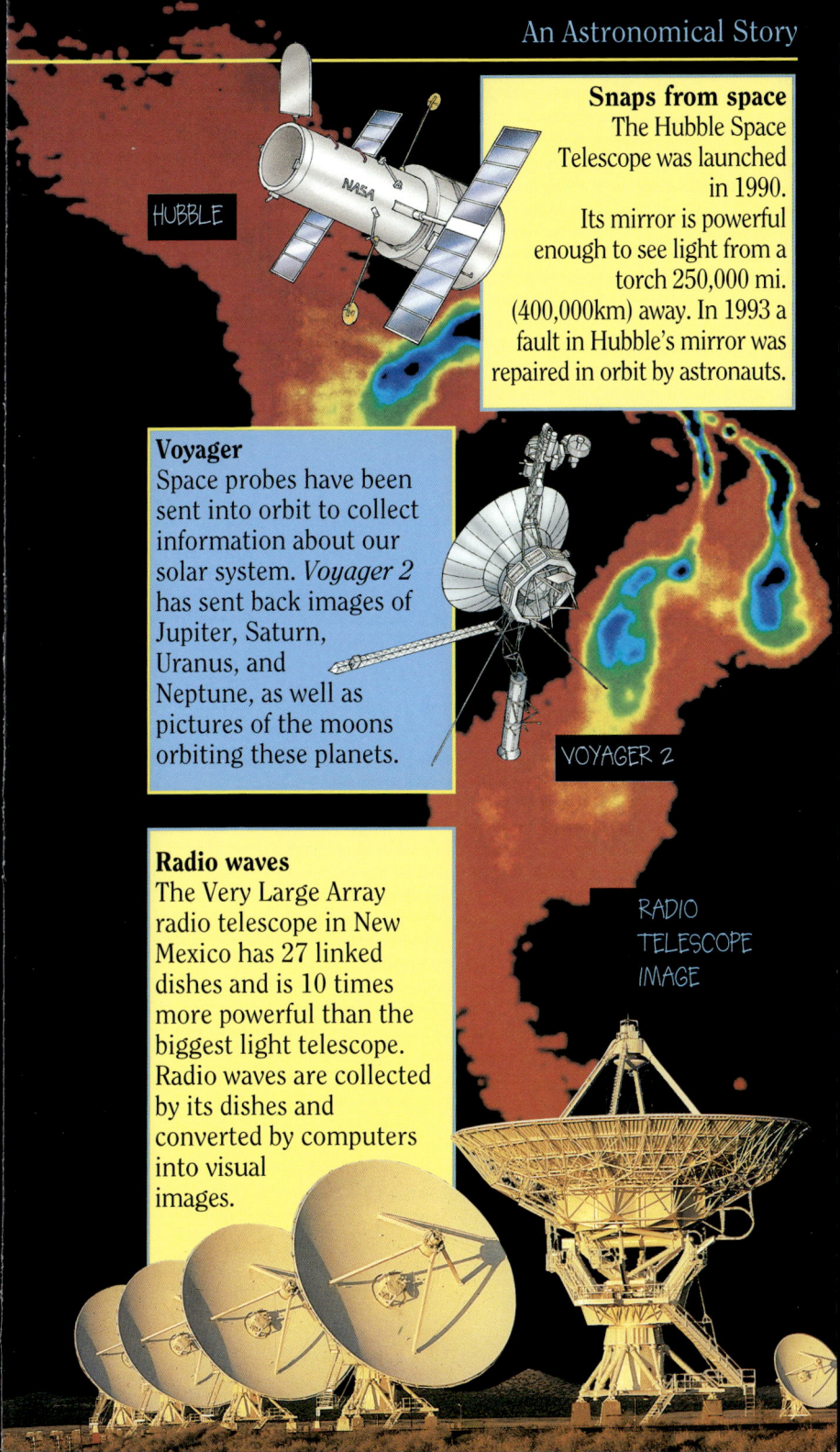

An Astronomical Story

Snaps from space
The Hubble Space Telescope was launched in 1990. Its mirror is powerful enough to see light from a torch 250,000 mi. (400,000km) away. In 1993 a fault in Hubble's mirror was repaired in orbit by astronauts.

HUBBLE

Voyager
Space probes have been sent into orbit to collect information about our solar system. *Voyager 2* has sent back images of Jupiter, Saturn, Uranus, and Neptune, as well as pictures of the moons orbiting these planets.

VOYAGER 2

Radio waves
The Very Large Array radio telescope in New Mexico has 27 linked dishes and is 10 times more powerful than the biggest light telescope. Radio waves are collected by its dishes and converted by computers into visual images.

RADIO TELESCOPE IMAGE

Blast-off!

The pull of Earth's gravity is so strong that rockets are needed to send probes and satellites through Earth's atmosphere and into space. Rockets burn liquid fuel to produce jets of hot exhaust gases.

Robert Goddard launched the first liquid fuel rocket in 1926.

The space shuttle

Until 1981 rockets could be used only once. They consisted of several engines and fuel tanks stacked on top of each other and each engine fell away as its fuel was used up. The shuttle has a revolutionary design that makes it reusable.

The boosters give the shuttle extra push for the lift-off. They drop off seconds into the flight and are recovered to be used again. The fuel tank holds 528,000 gallons (2 million liters) of fuel to power the orbiter's built-in rockets. It is the only part of the system that is not reusable.

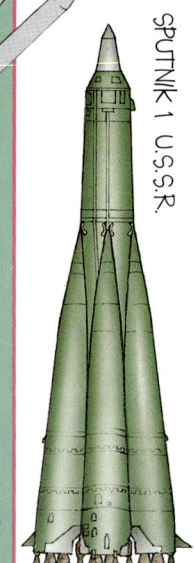

SPUTNIK 1 U.S.S.R.

First launch 1957

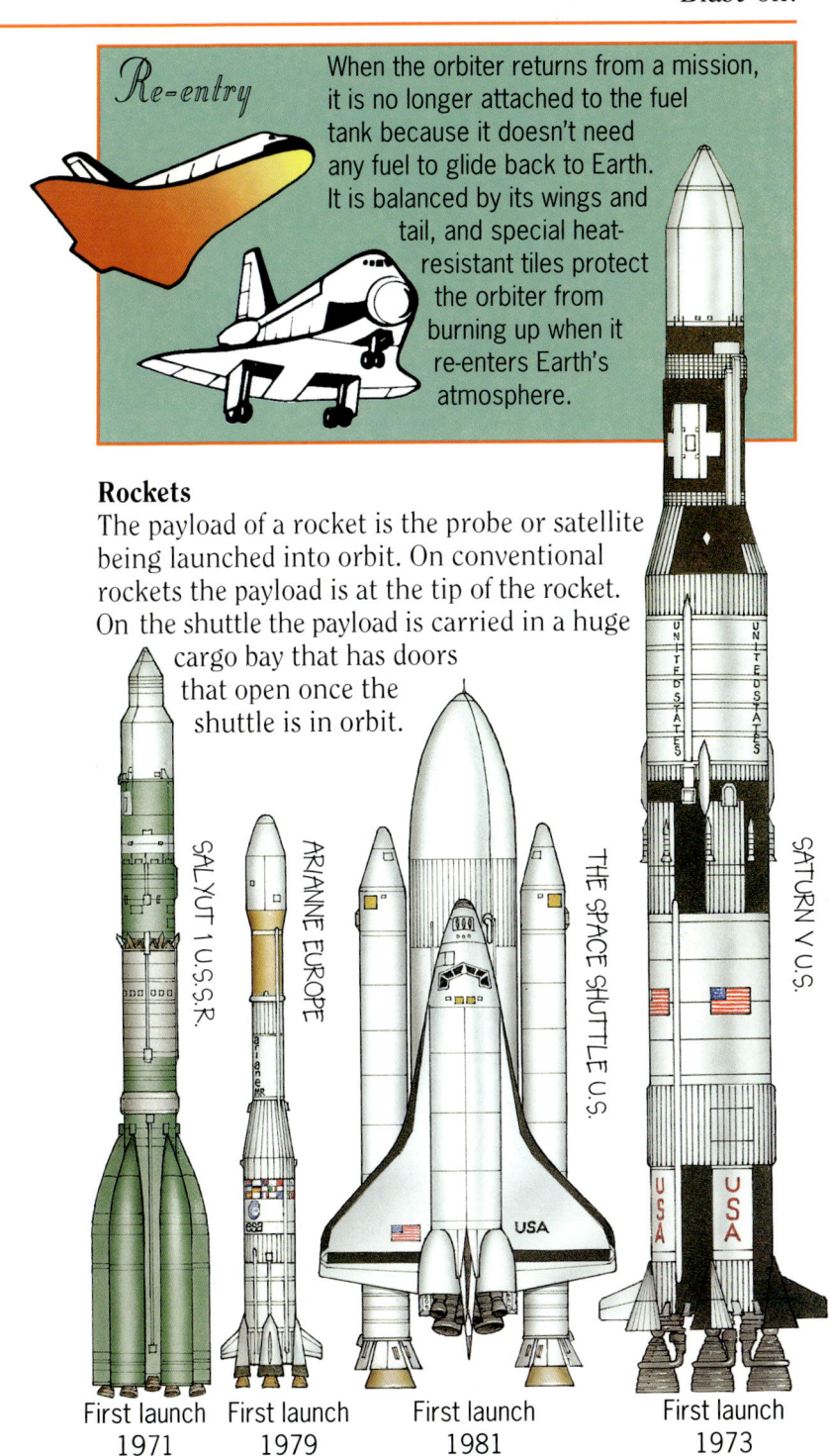

Blast-off!

Re-entry

When the orbiter returns from a mission, it is no longer attached to the fuel tank because it doesn't need any fuel to glide back to Earth. It is balanced by its wings and tail, and special heat-resistant tiles protect the orbiter from burning up when it re-enters Earth's atmosphere.

Rockets

The payload of a rocket is the probe or satellite being launched into orbit. On conventional rockets the payload is at the tip of the rocket. On the shuttle the payload is carried in a huge cargo bay that has doors that open once the shuttle is in orbit.

SALYUT 1 U.S.S.R.
First launch 1971

ARIANNE EUROPE
First launch 1979

THE SPACE SHUTTLE U.S.
First launch 1981

SATURN V U.S.
First launch 1973

Messages from Space

A satellite is any object that orbits another, so Earth is a satellite of the Sun. For millions of years Earth's only satellite was the Moon. Then, in 1957, *Sputnik 1* was launched and now our planet is orbited by hundreds of artificial satellites.

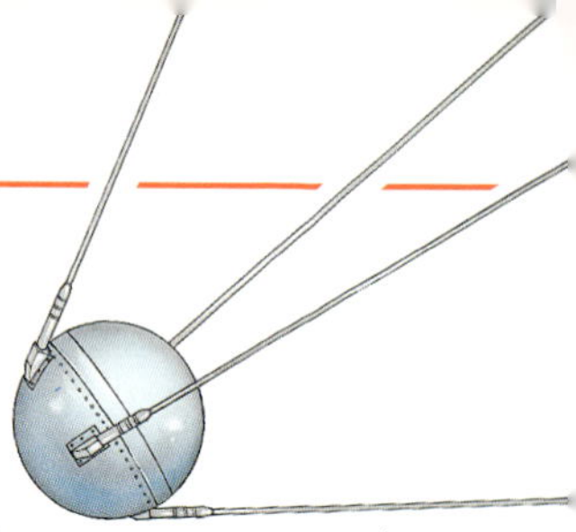

SPUTNIK 1

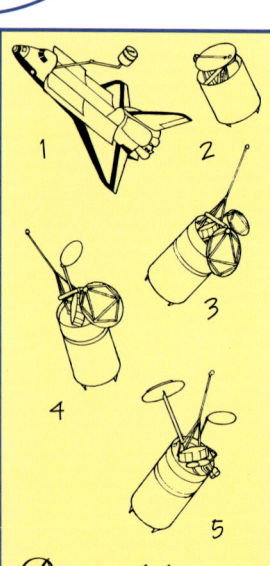

Weathersats
Meteorological satellites provide weather forecasters with regular pictures of global cloud cover to help them trace where storms begin and track their paths. They are also used to measure air and sea temperatures and wind speeds.

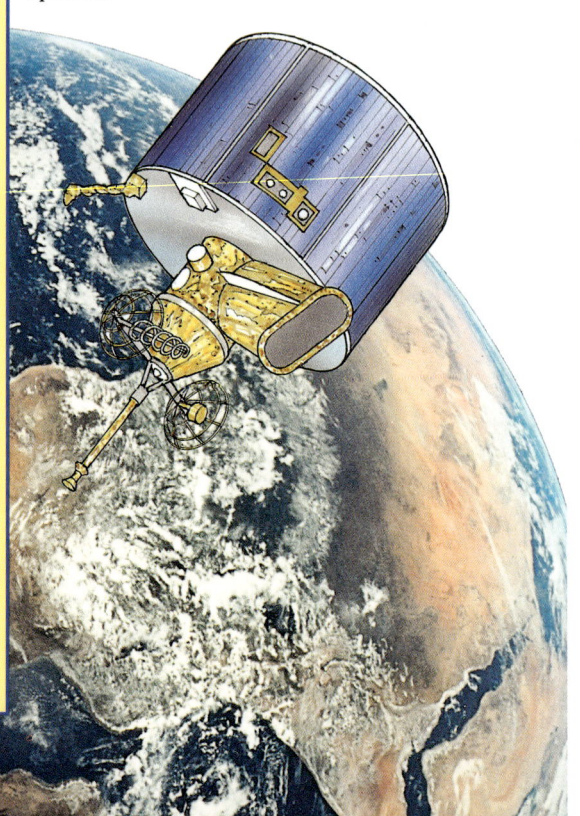

Powered by the sun
Satellites can be any shape or size but they all have solar panels to turn the Sun's energy into electricity. The solar panels, antennae, and beam reflectors unfold in stages once the satellite is put into orbit, as shown in the diagrams above.

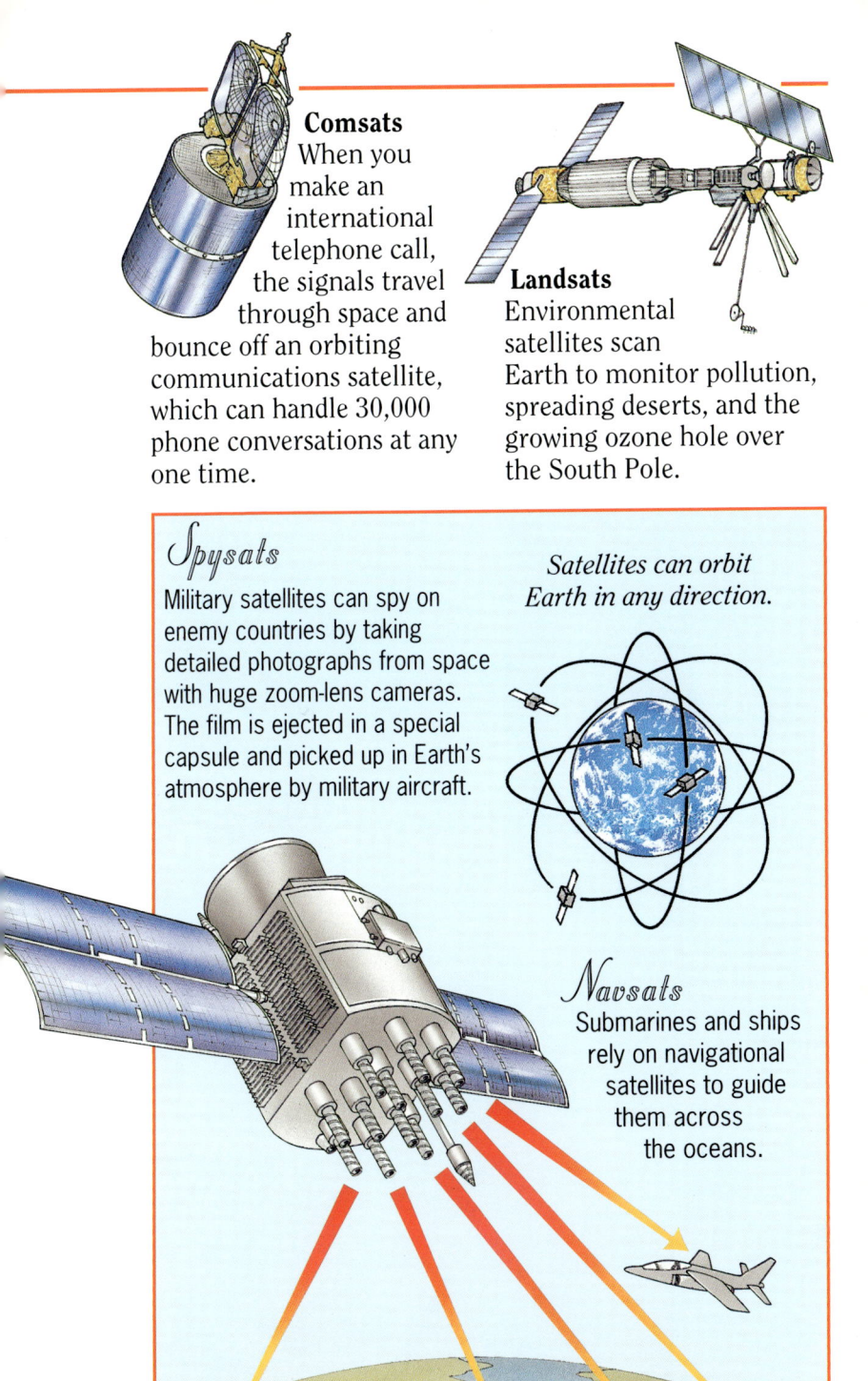

Comsats
When you make an international telephone call, the signals travel through space and bounce off an orbiting communications satellite, which can handle 30,000 phone conversations at any one time.

Landsats
Environmental satellites scan Earth to monitor pollution, spreading deserts, and the growing ozone hole over the South Pole.

Spysats
Military satellites can spy on enemy countries by taking detailed photographs from space with huge zoom-lens cameras. The film is ejected in a special capsule and picked up in Earth's atmosphere by military aircraft.

Satellites can orbit Earth in any direction.

Navsats
Submarines and ships rely on navigational satellites to guide them across the oceans.

Space Hotels

When Yuri Gagarin became the first space traveler in 1961, his orbit of Earth in the *Vostok 1* capsule lasted just 89 minutes. Now astronauts can stay in space for more than a year.

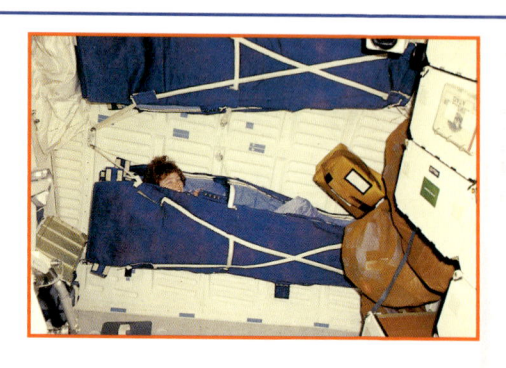

Tied to their beds!
In order to sleep, astronauts need to stop floating. They fix themselves into "sleep restraints," which are anchored to the walls of the spacecraft.

Wayward water
A shower in a spacecraft is different. Water on Earth falls down, but in space it floats like everything else, so a vacuum hose is used to suck up wayward drops.

Look out!
Without gravity, going to the bathroom in space could be hazardous! Astronauts have to use a special hose, which sucks their body wastes into a tank to be disinfected before being dumped overboard.

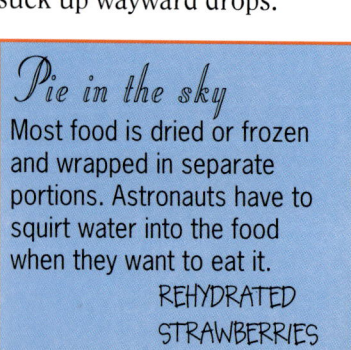

Pie in the sky

Most food is dried or frozen and wrapped in separate portions. Astronauts have to squirt water into the food when they want to eat it.

REHYDRATED STRAWBERRIES

Spacewalking
If astronauts have to fix the outside of a spacecraft, they strap themselves into a special armchair called a manned maneuvering unit. It is propelled by small jets of gas and allows them to move about in any direction for up to six hours.

Spacelabs

Space stations are sent into orbit so that crews of astronauts can visit them for periods of several weeks or even months to carry out scientific experiments. The first orbital space station, *Salyut*, was launched by the Russians in 1971.

SKYLAB

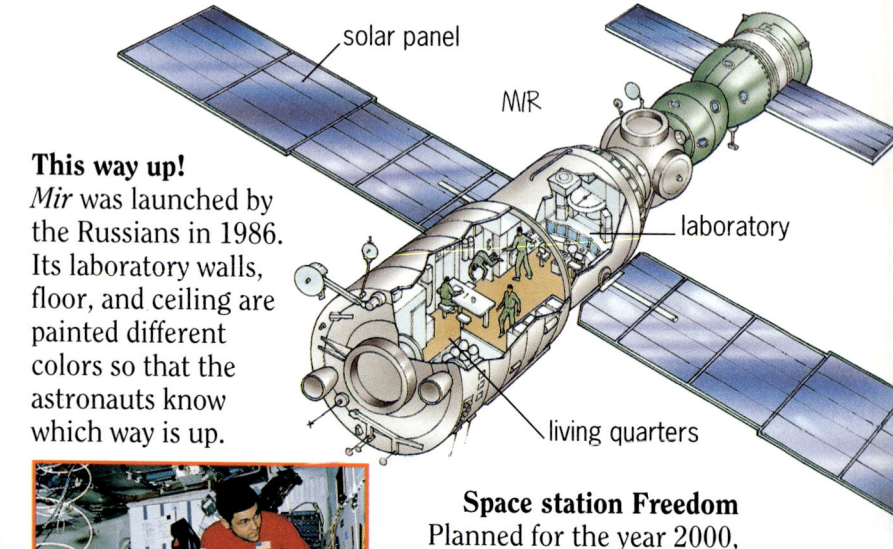

Recycled rocket
Skylab was launched in 1973 and stayed in space for five years. It was made out of an old rocket from one of the lunar missions and had separate living quarters below the laboratory for the visiting astronauts. In 1979 *Skylab* plunged out of control into Earth's atmosphere and burst into flames. Debris from the space station fell to Earth. Fortunately, its last mission had been completed in 1974.

solar panel

MIR

laboratory

living quarters

This way up!
Mir was launched by the Russians in 1986. Its laboratory walls, floor, and ceiling are painted different colors so that the astronauts know which way is up.

Space station Freedom
Planned for the year 2000, *Freedom* will be the biggest-ever, multinational, permanently manned, orbiting space station. It will be a repair shop for space vehicles and a launch pad for future missions.

Spacelabs

Did you know?
Astronauts grow about 2 in. (5cm) taller when they are in space because their bones are not pressed together by gravity. On long missions, crew members need to get regular exercise to keep their muscles toned.

Testing, testing

Oops!

Astronauts are themselves human guinea pigs while they are in space, but they have also made major scientific discoveries. In the weightless, germ-free conditions of space they have been able to make super-light metals and produce new medicines.

Future World

In the future, the Earth may no longer be able to support life. The deserts might spread or the salty oceans could flood the land. If this happens, people will have to leave the planet and live in orbiting space stations. These will need to have controlled atmospheric conditions so people can grow their own food. Perhaps some stations will travel beyond our solar system. Their crews may even encounter aliens from outer space and find themselves fighting star wars for their survival.

Future World

Cosmofacts

Mercury
Orbit = 88 Earth days
No moons
36 million mi. (58 million km) from the Sun

Venus
Orbit = 225 Earth days
No moons
67 million mi. (107 million km) from the Sun

Earth
Orbit = 365.25 days
1 moon
93 million mi. (149 million km) from the Sun

Mars
Orbit = 687 Earth days
2 moons
142 million mi. (227 million km) from the Sun

Jupiter
Orbit = 11.9 Earth years
16 moons
483 million mi. (773 million km) from the Sun

Saturn
Orbit = 29.5 Earth years
22 moons
887 million mi. (1.4 billion km) from the Sun

Eclipses

A solar eclipse occurs when the Moon moves directly in front of the Sun and blocks out the light. A total eclipse results in daytime darkness and can last for up to 7 minutes 31 seconds. The earliest recorded total solar eclipse dates back to 1217 B.C. and was seen in China.

Uranus
Orbit = 84 Earth years
15 moons
1.8 billion mi. (2.8 billion km) from the Sun

Neptune
Orbit = 165 Earth years
8 moons
2.8 billion mi. (4.4 billion km) from the Sun

Pluto
Orbit = 248.4 Earth years
1 moon
3.6 billion mi. (5.9 billion km) from the Sun

*Map of the solar system
Approximate distances of planets in the solar system from the Sun (not to scale).*

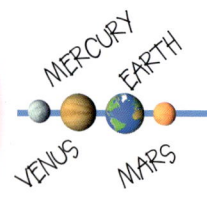

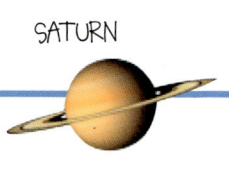

Cosmofacts

Gobbledegook
Shurnarkabtishashutu is the star with the longest name. This is an Arabic word meaning "under the southern horn of the bull."

Rag and bone
About four-fifths of the objects circling Earth are space junk like old rockets and satellites.

Take cover
Although no human has ever been killed by a meteorite, in Nahkla, Egypt, in 1911 a dog was not so lucky and died when it was hit by a meteorite.

Weight-watcher
Earth is putting on weight at a rate of 9,000 tons a year as meteorites and space dust fall to Earth.

Gigantic g-force
The force of gravity on the edges of a black hole is so incredibly huge that it would be like hanging from a bridge with the entire population of New York tied to your ankles.

Space firsts

Isaac Newton invented the mirror telescope in 1668.

Karl Jansky discovered radio waves from the galaxy in 1931.

First satellite, *Sputnik 1*, was launched in 1957.

First creature in space was a dog called Laika, sent into orbit in 1957.

First man in space was Yuri Gagarin in 1961.

First woman in space was Valentina Tereshkova in 1963.

INDEX

Andromeda Galaxy 7
Apollo spacecraft 11, 14
Ariane spacecraft 21
Armstrong, Neil 14
Asteroids 17
Astronauts 14–15, 24–27, 31
Astronomy 18–19
Bayeux Tapestry 17
Big Bang 2
Big Crunch 3
Black holes 9
Comets 16, 17
Copernicus, Nicolaus 18
Dinosaurs 3, 17
Earth 3, 4–5, 6, 13, 15, 17, 18, 22, 30
Eclipses 30–31
Freedom 26
Gagarin, Yuri 24, 31
Galaxies 3, 4
Galileo 18
Giotto spacecraft 17
Halley's comet 17
Herschel, William 18
Hubble Space Telescope 19
Jupiter 13, 19, 30
Light-years 6
Lowell, Percival 18
Luna spacecraft 15
Manned maneuvering unit 25
Mars 13, 18, 30
Martians 18
Mercury 13, 30
Meteorites 14, 16, 31
Meteoroids 16
Meteors 16
Milky Way Galaxy 3, 4
Mir 26
Moon 6, 14–15
Moon-buggy 14
Neptune 13, 19, 30
Newton, Isaac 31
Northern hemisphere 10, 13
Orion 7
Pluto 7, 13, 30
Polaris 7, 11
Proxima Centauri 7
Red giants 9
Rockets 20–21
Salyut spacecraft 21, 26
Satellites (artificial) 22–23, 31
Saturn 13, 19, 30
Saturn V spacecraft 21
Seasons 13
Skylab 26
Solar energy 9, 22
Solar flares 8
Solar system 4, 12–13, 17
Southern hemisphere 13
Space shuttle 7, 20, 21
Space stations 26
Sputnik spacecraft 20, 22, 31
Stars 2, 3, 7, 8–9, 10–11
Sun 4, 8–9, 12–13, 15, 16, 17, 18, 22, 30
Sunspots 8
Telescopes 18–19, 31
Tereshkova, Valentina 31
Tides 15
Universe 2–3
Uranus 13, 19, 30
Venus 13, 30
Voyager spacecraft 19
White dwarfs 9
Zodiac 10–11

Published in 1997 by Creative Education
123 South Broad Street
Mankato, Minnesota 56001
Creative Education is an imprint
of The Creative Company
Cover design by Eric Madsen
Illustrations by Julian Baum, Bill Donohoe,
Chris Orr, Charlotte Hard
Photographs by Michael Holford; NASA; Portfolio
Pictures/NASA; Robert Harding Picture Library;
Science Photo Library/François Gohier/Roger
Ressmeyer, Starlight/NRAD-AUI/NASA; Zefa

Text © HarperCollins Publishers Ltd. 1996
Illustrations © HarperCollins Publishers Ltd. 1996
Published by arrangement
with HarperCollins Publishers Ltd.
International copyrights reserved in all countries.
No part of this book may be reproduced in any form without written permission from the publisher.
Printed and bound in Hong Kong.

Library of Congress
Cataloging-in-Publication Data
Earth and space.
p. cm. - (It's a fact!)
Includes index.
Summary: Presents miscellaneous facts about the stars, planets, and other things whirling about in outer space.
ISBN 0-88682-859-7

1. Astronomy-Juvenile literature. 2. Outer space-Juvenile literature. 3. Earth-Juvenile literature.
[1. Astronomy-Miscellanea. 2. Outer space-Miscellanea.] I. Series.
QB46.E18 1997   96-35295
500.5-dc20
EDCBA